MW00470928

The LMH Official DICTIONARY of JAMAICAN HERBS & MEDICINAL PLANTS *and their uses*

Compiled by

L. Mike Henry / K. Sean Harris

First Edition
10 9 8 7 6 5 4 3 2 1

Second Edition
10 9 8 7 6 5 4 3 2
Reprint 2013
10 9 8 7 6 5 4 3

Compiled by: L. Mike Henry / K. Sean Harris
Cover Design: Susan Lee-Quee
Illustrator: Clovis Brown
Design/Typesetting: Michelle M. Mitchell

Published by: LMH Publishing Limited
Suite 10-11
Sagicor Industrial Park
7 Norman Road
Kingston CSO
Tel: 876-938-0005; Fax: 876-759-8752
Email: lmhbookpublishing@cwjamaica.com
Website: www.lmhpublishing.com

Printed in China

ISBN: 978-976-8184-33-7

Publisher's Notes

LMH Publishing is pleased to produce the first eleven books in a series of titles which will treat the Jamaican culture in a serious yet entertaining format.

The first eleven titles in this series are:

- LMH Official Dictionary of Jamaican Words & Proverbs
- LMH Official Dictionary of Popular Jamaican Phrases
- LMH Official Dictionary of Jamaican Herbs & Medicinal Plants and their uses.
- LMH Official Dictionary of Caribbean Herbs & Medicinal Plants and their uses.
- LMH Official Dictionary of Jamaican History
- LMH Official Dictionary of Sex Island Style
- LMH Official Dictionary of Sex Island Style Vol.2
- LMH Official Dictionary of Caribbean Exotic Fruits
- LMH Official Dictionary of Jamaican Religious Practices & Revival Cults
- LMH Official Dictionary of Bahamian Words & Proverbs
- LMH Official Dictionary of Popular Bahamian Phrases

As this series cannot be complete without the response of our readers (as no other publisher has yet attempted to record our culture) we implore you, our readers, to voice your opinions, comments and observations which we will take into consideration when publishing new editions.

I hope you enjoy our witty and innovative series. See you next time.

Mike Henry
Publisher

Disclaimer

The author(s) does not directly or indirectly diagnose, dispense medical advice or prescribe the use of herbs or any other substance as a form of treatment without medical approval. The intent is only to offer information. In the event you use this information without your doctor's approval, you are prescribing for yourself, which is your right, but the publisher and the author(s) assume no responsibility.

CONTENTS

INTRODUCTION

This book takes a look at some of the herbs and medicinal plants that are found in Jamaica. It also briefly explores some of the myths and legends associated with some of these herbs & medicinal plants.

What is a herb? A herb is any flower or plant with a succulent stem which dies at the root every year and is used in medicine or cooking. Herbs have been used for centuries as both food and medicine. Many modern drugs are originally derived from herbal sources.

Most herbs have a specific application and should be treated as medicine; utilized only for the prescribed ailment. Caution should be exercised when prescribing herbs. It is best to know the medical history of the patient and the history of the herb, before taking or prescribing any herbal remedy.

Herbs allow us to maintain our natural state of good health. One should use herbs wisely, moderately and regularly. The healing properties of herbs are of a wide and diverse nature, so it is easy for one to incorporate the use of herbs in one's daily life in order to attain and maintain a healthy lifestyle.

L. Mike Henry /
K. Sean Harris

Jamaican Herbs

ACKEE (Blighia Sapida)

Ackee is edible, and in Jamaica it is consumed with salt fish as the national dish. However, the fruit must be allowed to ripen on the tree, otherwise it could be poisonous.

Medicinal and other uses:

a. The leaves are used to make a tea for colds, flu, asthma and mucus congestion.

b. The tea can also be mixed with salt and used as a mouthwash for pyorrhoea and gum problems.

c. Leaves hung in bunches on porches and around the house will keep away flies and other insects.

d. The skin of the ackee can be beaten to a pulp and used as a kind of soap to wash cloth.

Ackee

ALOE VERA

Aloe vera is very strengthening and cleansing to the entire body. Aloe is rich in calcium, potassium, and vitamin B-12.

Medicinal and other uses:

a. It cleanses the kidneys, bladder and removes all morbid matter from the stomach and intestines.

b. It is a strong laxative that cleans and purifies the intestines, bowels and colon.

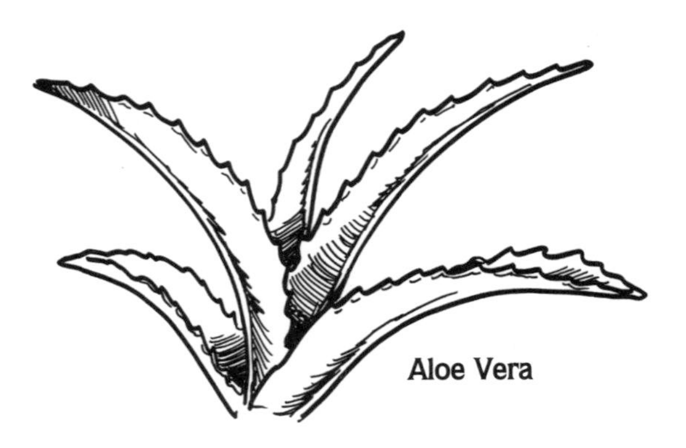

Aloe Vera

c. The dried gel of the aloe vera leaves are mixed with unsweetened natural wine and used as a medicine for diabetes.

d. It clears up red eyes, improves vision and brings out the natural colour of the eyes.

e. It is used for sunburn, skin irritations and insect bites.

f. It removes wrinkles and other discolorations from the skin.

g. When mixed with lemon juice and applied as a poultice it heals skin ulcers, boils and ringworm.

h. Aloe also relieves dandruff, stimulates hair growth, strengthens the hair and kills bacteria associated with scalp disease.

i. It heals and soothes burns, plus it prevents blistering and discolouration.

AVOCADO (Persea Americana)

Avocadoes are rich in protein, vitamin A, B, C, D, E, and chlorophyll, calcium, potassium, phosphorus and iron. The fatty acid in avocado is highly digestible. In Jamaica, avocadoes are usually eaten with lunch or dinner.

Medicinal and other uses:

a. A tea of the leaves or bark is used for colds, coughs, asthma and high blood pressure.

b. The leather branches can also be boiled and used for diarrhea, hypertension and pains.

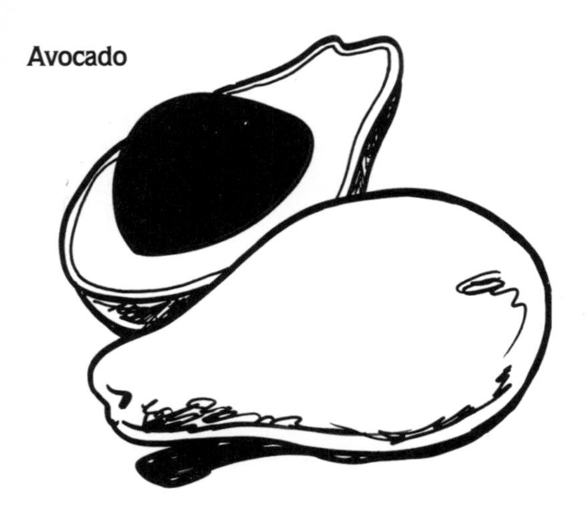

c. Avocado seeds contain antibiotic substances that will kill rats when mixed with cheese.

d. Heated leaves applied to the head relieve headaches.

BLACK SAGE (Cordia Globosa)

Medicinal and other uses:

a. In Jamaica, the tea is used for menstrual cramps, colds, gout, mucus congestion and tightness in the chest.

b. The tea is also used for internal bleeding and as a bath for skin eruptions.

BREADFRUIT LEAVES
(Artocarpus Altilis)

The breadfruit tree has a wide range of uses. The fruit is edible and may be prepared in a number of ways, and the leaves and the sap which runs from the tree can be used for various ailments.

Medicinal and other uses:

a. A tea of breadfruit leaves, which contain camphoral, is used to lower blood pressure and treat diabetes.

b. In Jamaica, the sap is applied to contagious skin ailments to prevent their spreading and promote healing.

c. The sap can also be boiled to form a rubbery substance that is used to dress wounds.

BRIDES TEARS
(Antigonon Leptopus)

The tubers at the root and the flower clusters can be cooked and eaten as food.

Medicinal and other uses:

a. The leaves and flowers are used as a tea for coughs, colds and constriction of the throat.

BROOM WEED
(Spartium Scoparius)

The active ingredient in this herb is spartien sulfate.

Medicinal and other uses:

a. The tea of this herb strengthens the heart, slows it down and reduces blood pressure.

b. It also increases the flow of urine and is effective for treating edema.

CALABASH (Crescentia Cujete)

In Jamaica, the dried shells are used as natural bowls for soups and water.

Calabash

Medicinal and other uses:

a. The pulp of the green fruit mixed with the leaves of the guaciacum or lignum vitae is taken for diabetes.

b. The young fruit is warmed, split open and applied to abscesses and boils to draw out infection and bring them down.

CANNABIS (Indica, Cannabis Sativa, Sativa Ruberlis)

Cannabis

Some common names for this popular herb are marijuana, weed, reefer, pot and ganja.

Medicinal and other uses:

a. Cannabis is used as a liniment for sprains, arthritic pains, asthma, bronchitis and migraine headaches.

b. A poultice can be made for sores, external ulcers, sprains, gout, boils, arthritis, infections and to kill pain.

c. Fresh leaves can be crushed, lightly heated and applied to sores, ulcers, boils, swellings, sprains and pains.

d. Cannabis will stimulate hair growth and get rid of dandruff and other scalp problems. It can be used as a final rinse or tea rubbed into the scalp.

e. Can be used as a seasoning as the seeds contain a very nutritious oil and can be sprinkled in soups and other dishes.

CASHEW
(Anacardium accidentale)

The nuts are edible and are quite tasty. Perfect for cocktails.

Medicinal and other uses:

a. A tea made of the bark and leaves can be taken for diarrhea, indigestion, stomach pains, bowel disorders and dysentery.

b. The oil of the seeds of the cashew removes warts, corns and freckles.

c. The shells soaked in water make a good insecticide that can spray on plants.

CASSAVA (Manihot Esculenta)

There are two types of cassava, the bitter and the sweet. Both are widely used and are edible. Cassava leaves can be steamed and eaten as a vegetable.

Medicinal and other uses:

a. Freshly cut cassava can be applied to snakebites, boils and external ulcers to promote healing.

b. It can also be grated and applied as a poultice for eczema and abscesses.

c. The fresh leaves applied to the head will relieve headaches.

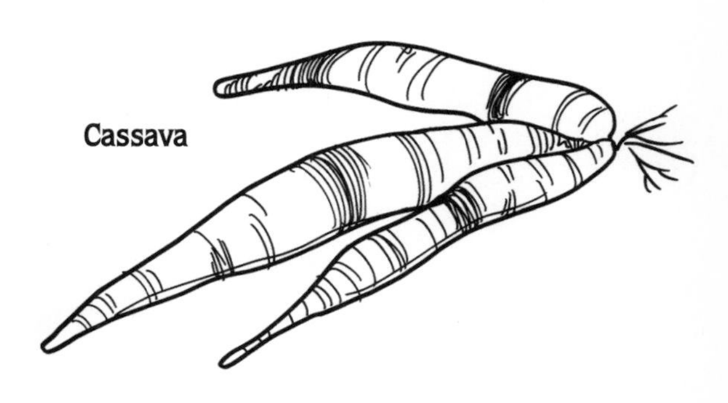

CASTOR OIL PLANT
(Ricinus Communis)

Castor oil has a very radical effect on the stomach and intestines and is not recommended for internal use in large quantities especially for children.

Medicinal and other uses:

a. Warm castor oil is used as a rub for arthritis and as a bath oil it relieves many skin problems.

b. The plant is also used to cure constipation and bodily pain.

c. Hot castor oil massaged into the hair will strengthen it and eliminate dandruff, lice, fleas and other scalp problems.

CERASEE
(Momordica Charantia)

Cerasee can be cultivated as a vegetable and cooked like other leafy vegetables. The green, unripe fruit can be soaked in water and cooked in curry and other dishes.

Medicinal and other uses:

a. Cerasee tea is used as a remedy for colitis, liver complaints, fever and as a skin lotion.

b. The juice of the ripe fruit, which contains valuable enzymes and minerals, is taken for diabetes.

c. A cerasee bath is good for arthritis, gout and other ailments.

CHAMOMILE (Anthemis Nobilis)

Some common names for chamomile are manzanilla and ground apple. Chamomile is one of the most widely used herbs in the world.

Medicinal and other uses:

a. It is one of the best known tonics to strengthen a weak stomach, relieve cramps, gas heartburn and aid digestion.

b. The tea relaxes the nerves and relieves muscle spasms.

c. The tea makes an excellent eyewash for burning and tired eyes.

d. Holding chamomile in the mouth will relieve toothaches and other mouth pains.

e. As a poultice, it reduces swelling and prevents gangrene.

f. It is also a tonic for all the female organs. It regulates and brings on menstrual flow and relieves uterine cramps.

g. Applied externally, the tea removes puffiness and black circles from around the eyes.
h. The tea rubbed on the skin will keep away insects, especially mosquitoes.

CHANEY ROOT (Smilax Havanensis)

Medicinal and other uses:

a. The root is used as a blood cleanser which is good for all skin problems.
b. It is used as a treatment for syphilis, rheumatism and arthritis.
c. The fruit is used to make a tonic wine.

COCOA (Theobrama Cacao)

The ripe fruit of the cacao tree is very sweet and delicious and contains life-giving enzymes and minerals.

Medicinal and other uses:

a. In Jamaica,

cocoa tea is used to stimulate the central nervous system, dilates blood vessels thereby lowering blood pressure.

b. The cocoa oil, known as cocoa butter can be used on cracked lips, burns, sore breasts and genitals as well as in the vagina and rectum to relieve irritations.

c. Cocoa butter when applied to minor keloids and scars reduces same.

COCONUT (Cocus Nucifera)

Coconuts are a staple food in most Caribbean islands. Both the milk and oil are used extensively in cooking. Coconut water has no impurities because it is collected one drop at a time and totally filtered by the fibrous roots and trunk of the plant.

Medicinal and other uses:

a. Coconut water is diuretic and flushes the kidneys, bladder and urethra.

b. Coconut milk is used externally on swollen and inflamed breasts caused by nursing.

c. Coconut oil can be given to babies to relieve choking and throat irritations.

d. A tea made of the roots will ease toothache, reduce body heat, stop menstrual hemorrhaging and help cure venereal disease.

e. As a skin oil, it rids the body of lice, rashes and other skin problems.

f. A decoction made from boiling the fibres of the mature husk is effective against diarrhea and thrush.

g. Coconut oil is also used for making natural soaps and other health products.

h. The very immature nuts that often fall prematurely can be boiled with similax and other roots to make tonics.

COFFEE (Coffea, Arabia)

Medicinal and other uses:

a. Coffee is a brain stimulant.

b. Its also used for narcotic poisoning, snake bites and to help ward off a coma.

c. A tea of fresh unroasted seeds can be taken for fever and jaundice.

COLA NUT (Sterculia)

The seeds of the cola nut contain 46% starch, 2.36% caffeine, 3% fat and small amounts of glucose and sucrose. In Jamaica, a tea of the grated cola nut is used as a coffee. The seeds are chewed as a stimulant because of their natural caffeine content.

Medicinal and other uses:

a. A tea of the cola nut regulates circulation and cardiac rhythms, and acts diuretically on the heart.

b. It improves the appetite and lessens the desire for alcohol.

c. The plant produces natural eye drops and this liquid clears up pink eye and removes cataracts and other growths in the eye.

COMFREY (Symphytum Officinale)

Some common names for comfrey are Knitback, wallwort, blackwort and salsify. Comfrey' s ability to heal comes from the element allantoin, which is a cell proliferant, which speeds up the body's natural healing process.

Medicinal and other uses:

a. The tea applied to the skin will heal all irritations and rashes.

b. A poultice of the leaves or roots applied to boils, swellings, and sprains will heal and draw out impurities.

c. Used as a mouthwash the tea will stop bleeding gums and relieve sore throats.

d. Comfrey is one of the best remedies for asthma, ulcers, diseased lungs and all bronchial problems.

DANDELION (Taraxacum Offcinale)

Dandelions are one of the oldest herbal remedies known to man and it's also strange in that it only thrives where there are human inhabitants. The flowers are rich in vitamin C and D as well as trace minerals and can be eaten raw or made into tonic wine. The roasted ground roots can be used as a caffeine-free coffee substitute.

Medicinal and other uses:

a. Dandelion roots are one of the best herbs for obstructions of the spleen, liver and bladder.

b. Dandelions are also used to treat jaundice, eczema and psoriasis.

c. Dandelions are very diuretic and help to remove excess mucus, strengthen the bones, add firmness to the teeth and help to prevent pyorrhoea.

d. Dandelions are an effective treatment for diabetes since they contain natural iodine, which aids the pancreas in excreting normal amounts of insulin.

DEVIL'S HORSE WHIP (Achyranthes)

Medicinal and other uses:

a. Used in the West Indies for coughs, colds, chest pains, colic, fever and venereal disease.

b. A tea of the leaves and roots is good for colds in infants.

DILL (Anethum Graveolens)

Medicinal and other uses:

a. Dill seeds and weed can be made into a soothing tea to relieve stomach aches and colic in restless infants.

b. Dill is also a good remedy for hiccups.

DONKEY WEED (Stylosanthes Hamata)

Some common names for donkey weed are cheesy toes, lady fingers and percil flowers.

Medicinal and other uses:

a. The sap is used to remove warts, sores, and moles.

b. A tea is used for kidney pains, colds, teething in babies and for reducing fevers.

DUPPY GUN (Reullla Tuberosa)

Medicinal and other uses:

a. A tea of the roots and leaves is good for coughs, colds and inflammation of the stomach and intestines.

b. The roots can be used for oligura, flu, constipation and venereal disease.

EUCALYPTUS (Eucalptus Globulus)

Medicinal and other uses:

a. A tea of the leaves is used for asthma, bronchitis and diabetes.

b. The leaves can be simmered in an open pot and the vapours inhaled for sinus, hay fever, colds and chest congestion.

c. The tea is also used as an antiseptic dressing for wounds, sores and open ulcers.

d. The fresh leaves can be chewed to strengthen teeth and harden the gums.

FEVER GRASS (Andropogon Citratus)

The chilled tea makes a refreshing beverage, perfect for hot weather.

Medicinal and other uses:

a. The roots are made into a tea and used as a mouthwash for gum problems and pyorrhoea.

b. The tea is also used for colds, fevers, malaria, flu, coughs, pneumonia and head-aches.

c. The roots can also be chewed until frayed out to form a natural tooth-brush.

FIT WEED (Eryngulm Foltidum)

Fit weed is rich in iron, calcium, carotene and riboflavin.

Medicinal and other uses:

a. In Jamaica, the tea is used to stop convulsions and is rubbed on the body in the case of fainting fits.

b. It is also used to stop uterine hemorrhaging, to relieve gas, and all stomach problems.

c. Fit weed is also used as a seasoning for fish and poultry.

FLAX (Linum Usitatissimum)

Flax, commonly known as linseed, is rich in chloride, sulphate, phosphate, calcium, potassium and magnesium.

Medicinal and other uses:

a. A decoction made of the seeds, are used for lung and chest problems, urinary tract disorders, coughs, catarrh, asthma, dysentery and to relieve menstrual cramps.

b. The fresh leaves can be used as a poultice for swellings.

c. A poultice can also be made of the powdered seed, along with warm water. This can be used to draw out boils, tumors, carbuncles, sores and abscesses.

d. Linseed oil, which is extracted from the seeds, is used as a laxative and also to relieve coughs, asthma and pleurisy.

GARLIC

This herb has excellent cleansing action.

Medicinal and other uses:

a. Garlic is used in the treatment of athlete's foot.

b. It is also used to control high blood pressure and cholesterol.

c. Garlic is also used as a tonic and cough suppressant.

GINGER (Zingiber Officinale)

Ginger is an appetite stimulant and can be added to various dishes to make them more digestible.

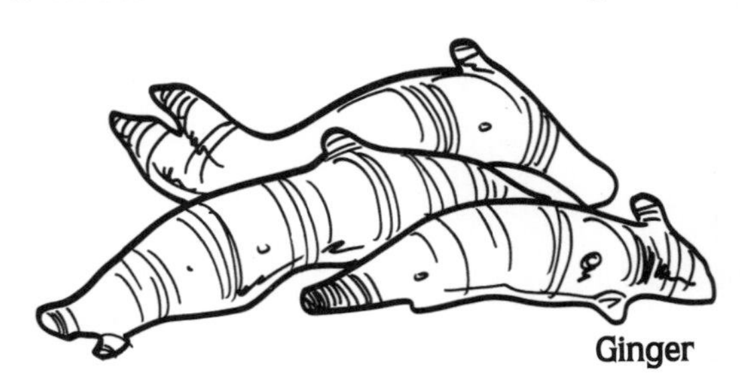

Ginger

Medicinal and other uses:

a. Ginger warms the body and helps to eliminate wastes through the skin.

b. The fresh roots can be chewed to relieve headaches.

c. Fresh or dried ginger is used as a tea to stop the production of mucus and fat in the body as well as for asthma, coughs, colds, flu and gastric problems.

d. Powdered ginger is made into a paste and spread on brown paper and applied to the forehead to relieve headaches.

e. The root expels gas, and controls high blood pressure and high cholesterol.

GOD BUSH (Oryctanthus Occidentalis)

God bush also known as 'Scorn the Earth', grows in trees and when picking one should not use metal scissors etc, or allow the bush to touch the ground, as this will render it weak and useless.

Medicinal and other uses:

a. God bush is used in Jamaica for female womb problems.

b. The tea of this plant is used for nervous problems, insomnia, fever and high blood pressure.

GOAT CORN (Morinda Royac)

Goat corn is also known by the names stiff cock, yellow ginger, strong back and duppy poison.

Medicinal and other uses:

a. In Jamaica, goat corn is used by itself or mixed with sarsaparilla root and young coconut, as a blood and nerve tonic.

b. A decoction of the plant, stem and leaves is used as a bath for mental derangement.

HONEY

Honey is the best sweetener known to man. The natural sugars are already broken down, making it easily assimilated into the blood stream. It contains all known vitamins and minerals, depending on the type of honey and the method used in producing and bottling.

Medicinal and other uses:

a. When pure honey is applied to burns, it will prevent infections and stop bleeding.

b. Honey is one of the best cough medicines; it will break up mucus and phlegm.

c. Honey is good for chronic constipation, having a menstrual laxative effect.

HORSERADISH
(Armoracia Lapathifolia)

Medicinal and other uses:

a. Fresh horseradish grated and mixed with a little honey can be taken by the spoonful to relieve asthma attacks, colds, coughs and mucous congestion in the chest.

b. Horseradish contains enzymes which when mixed with hydrogen peroxide, removes chemical wastes and pollutants from water.

IRISH MOSS
(Chondrus Crispus)

Some common names for Irish moss are sea moss and carragheen. Irish moss contains all minerals needed by the body and makes a good mineral supplement.

Medicinal and other uses:

a. Irish moss is rich in iodine, making it beneficial for strengthening, nourishing and toning up the glands.

b. When mixed with flaxseeds, it is good for all urinary tract ailments.

JACK IN THE BUSH
(Eupatorium Odoratum)

Medicinal and other uses:

a. In the Caribbean a tea made of the leaves is used for colds, flu and fever.

b. A tea of the flowers is used for coughs and diabetes.

c. A tea of the shoots can be mixed with coconut milk for bronchitis in children.

JACK FRUIT
(Artocarpus Heterophyllus)

Jack fruit contains protein, fat, calcium, phosphorus and iron in quantities normally present in other fruits. The pulp is rich in sugars; a fair amount of carotene is also present, but it is low in vitamin C.

Medicinal and other uses:

a. The unripe fruit is used for blood clotting.

b. A tea is made from the bark and drunk as cooling.

c. The ripe fruit is eaten raw as a laxative and cooling.

d. The unripe fruits are used as a vegetable, in soups and made into pickles, syrup, jam, jelly and candy.

Jackfruit

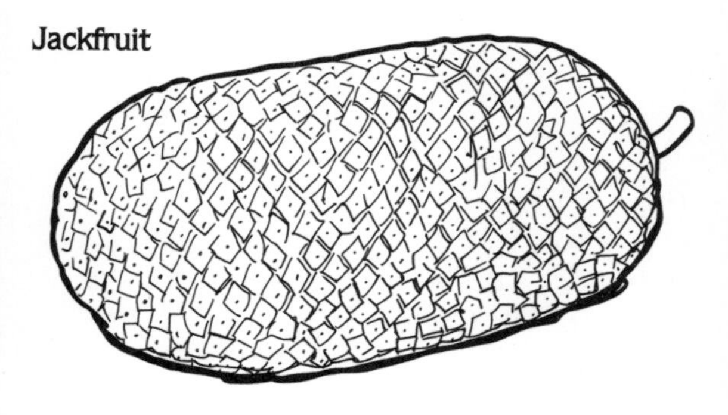

JAMAICAN CHEWSTICKS (Gouania Lupilodes)

Medicinal and other uses:

a. When used as a toothpaste, it will remove tart-tar, kills bacteria that causes tooth decay, stop bleeding gums and tighten loose teeth.

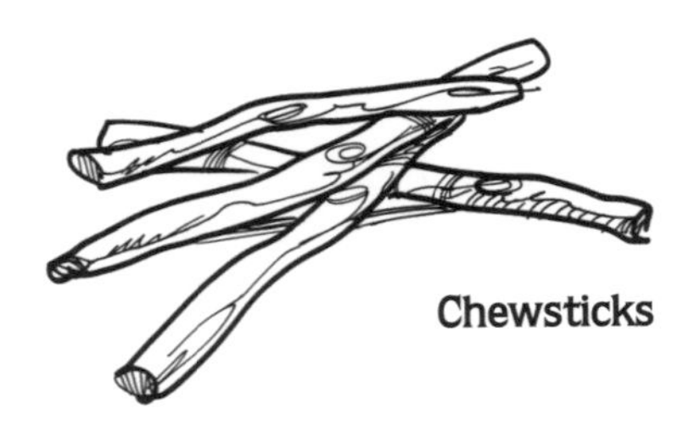

Chewsticks

KHUS KHUS (Vetiveria Zizanioides)

Medicinal and other uses:

a. Freshens the air and keeps away insects.

b. When placed in closets and drawers will keep away moths and sweeten clothes.

KOLA NUT

A common name for Kola Nut is bissy. An extract from Kola nut is used all around the world in the making of kola drinks (kolas). There are 125 species of the plant.

Medicinal and other uses:

a. The powder from the Kola Nut aids in healing cuts and bruises.

b. In Jamaica, Kola Nut is used as an antidote against poison.

c. Kola nut is used in small doses for migraine, motion and morning sickness.

d. The masticated Kola Nut is used to allay hunger, prevent thirst, open the appetite and sustain strength.

LEAF OF LIFE (Bryophyllum Pinatum)

Leaf of life is also known as love plant, love bush, live forever and life plant.

Medicinal and other uses:

a. A tea of this herb is used for shortness of breath, kidney failure, menstrual problems, asthma, and chest colds.

b. It also removes harmful bacteria from the intestines and clears the urine.

c. A poultice of the crushed leaves can be applied to body for sprains, pains and earaches.

d. The crushed leaves also relieves insect bites, bruises, boils and skin ulcers.

e. The fresh leaves can be eaten raw for asthma, bronchitis and intestinal problems.

LEMON LEAF

Medicinal and other uses:

a. A tea of the flower buds is used to sweeten up the system.

b. It is a good disinfectant for wounds and is used for liver complaints as well as fevers.

c. The juice of the lime is mixed with water to stop diarrhea and it is flushed up the nostrils to stop nosebleeds.

LEMON GRASS (Cymbopogon Citratus)

Medicinal and other uses:

a. In the Caribbean, it is used to expel gas, relieve colic, and stomach pains.

b. The oils in lemongrass make it a natural insect repellent that can be rubbed on the body.

c. The tea sweetens up the entire system, making perspiration less offensive by acting as a natural deodorant.

LIGNUM VITAE
(Guacicum Officinale)

The wood is one of the hardest in the world and was once used for shipbuilding as well as medicine.

Medicinal and other uses:

a. The inner bark is made into a liniment for sprains and pains due to arthritis.

b. The powdered wood can be packed into cavities for a few days to loosen teeth, making extraction painless, with little or no bleeding.

c. A tea of the leaves is also used as a wash to strengthen the eyes.

LILY OF THE VALLEY
(Convallaria Majalis)

Medicinal and other uses:

a. A good heart tonic; it increases energy and regulates the heartbeat.

LOVE BUSH

Medicinal and other uses:

a. Love bush is used for stomach problems, griping in babies, nerves and as a lotion for prickly heat rash.

MANGO
(Mangifera Indica)

Mangoes are an important source of vitamin A, fair in vitamin B, and with varying quantities of vitamin C. There are many different species of mango.

Medicinal and other uses:

a. A tea of the leaves can be used to stop diarrhea, fever and to prevent insomnia and hypertension.

b. A tea of the mango skins can be used as a wash for infected sores and external ulcers.

c. The gum or resin of the tree can be made into a poultice for colds and chest congestion.

MAN TO MAN BUSH

Medicinal and other uses:

a. In Jamaica, the juice is dropped into the eyes to relieve the pressure of glaucoma.

MAN PIABA BUSH

Medicinal and other uses:

a. The tea is used to strengthen the male reproductive organs.

MARIGOLD
(Calenduia Officinalis)

Medicinal and other uses:

a. As a poultice, the flowers are used externally to bring down swellings, heal sores, and draw out boils and abscesses.

b. Rubbing the fresh flowers on the body will remove scars, warts and other bodily marks.

c. The tea is also used to relieve pains due to cancer and is useful in rebuilding cells rapidly.

d. A tea of the fresh flowers can be used as a rinse for blonde hair to bring out highlights.

e. Marigolds serve as a useful insecticide, keeping away most harmful insect pests.

f. The dried flowers are used to colour soups, gravies and dairy products.

NEASEBERRY
(Manilkara Sapota)

Medicinal and other uses:

a. In Jamaica, the leaves are used as a nerve tonic.

b. The crushed seeds mixed in water can be used as an antidote to rabies and can be applied to all poisonous insect or animal bites.
c. The sap, known as chicle, is used as a filling for decayed teeth and is also chewed like gum.
d. The yellow leaves are used for colds, flu, fevers, coughs and diarrhea.

NETTLE (Urtica Dioica)

Nettle grows wild and abundantly in many countries. The entire plant is covered with fine needles, that sting when you touch them. Nettle is rich in vitamin A, calcium and other vitamins and minerals.

Medicinal and other uses:

a. Nettle tea or the fresh juice will promote the flow of breast milk and will also increase menstrual flow.
b. The fresh boiled leaves will stop bleeding and promote healing.
c. Nettle is also good for those suffering from piles and hemorrhoids.
d. Nettle can be applied to the skin or taken internally as a remedy for rashes and pimples.
e. Nettle is very effective for sunburn and will clean

the skin and improve circulation, giving the skin a healthy glow.

f. As a hair rinse, it will bring back natural colour to the hair and rid the scalp of dandruff as well as promoting hair growth.

NUTMEG (Myristicaceae Fragans)

The nutmeg tree takes nine years to bear fruit and only the female bears fruit. The best fruit comes from trees near the sea. The seeds of the fruit are dried in the sun until the kernel starts to rattle after which the shell is broken and the nutmeg is extracted.

Medicinal and other uses:

a. In the Caribbean, nutmeg is used for asthma, malaria, fever, coughs and diarrhea.

b. A lotion can be made with nutmeg for arthritis and painful joints.

c. Nutmeg stuffed into cavities, relieves toothaches.

d. Holding a piece of nutmeg in the mouth will freshen your breath.

OBEAH BUSH

Obeah bush is also known as granny bush, worry bush and goat weed.

Medicinal and other uses:

a. A tea of this herb can be taken to hasten child-birth and lessen labour pains.

b. It also relieves indigestion in pregnant women.

OBI SEEDS (Bissy)

Medicinal and other uses:

a. Used in the Caribbean for blood pressure, heart problems, diabetes and poisoning.

PAK CHOI (Brassica Chinensis)

The leaves contain a high percentage of iron, phosphorus and calories.

Medicinal and other uses:

a. The boiled water of the pak choi is considered as a tonic.

b. The leaves are edible and are used in the making of various dishes.

PARSLEY (Petroselinum Sativum)

There are many species of parsley and they all have similar properties. Parsley contains vitamin D, E, and B, calcium, sulphur, potassium and phosphorus.

Medicinal and other uses:

a. Parsley is effective in treating all ailments of the optic nerves such as weak eyes, cataracts and glaucoma.

b. A tea of the roots and the dried leaves is used to revitalize the thyroid gland and has a cleansing effect on the kidneys, bladder and urinary system.

c. Fresh parsley placed on the eyes will remove black bags and puffiness.

d. Parsley is ideal for weight loss since it appeases the appetite and helps to dissolve fat.

e. Fresh sprigs of parsley eaten after a meal will sweeten the breath.

f. As a hair rinse it makes the hair soft and glossy.

g. Parsley is very beneficial to those suffering from low blood sugar.

h. Parsley planted around rosebushes will eliminate rose beetles.

PASSIONFLOWER (Passifora Incarnata)

The flower of the plant, which is really a vine, is purple, white, blue or pink and has a cross in the center with a crown of feathery rays.

Medicinal and other uses:

a. Passionflower is one of the best herbal nerve tonics known and is often the main ingredient in many sedatives.

b. It soothes and calms the nerves without bringing on depression or affecting one's ability to concentrate.

c. It also improves the circulation, lowers blood pressure and cures insomnia.

PEPPER ELDER

Some common names for pepper elder are black joint, Jamaican black pepper and jointwood.

Medicinal and other uses:

a. The stems and leaves of this plant are used as a tea which is mixed with ginger and taken for all stomach problems, especially cramps and pains.

b. It is also used for constipation, fever, as a blood purifier and a natural sedative.

c. The leaves and stems can be used in an herbal bath to relieve bodily pains, break up chest colds and mucous congestion in the bronchial passages.

d. In Jamaica, the fruit is used as a substitute for black pepper.

PEPPERMINT

A common name for peppermint is Jamaican Fine leaf Peppermint.

Medicinal and other uses:

a. In Jamaica, a tea made of this mint is used to strengthen the stomach, relieve cramps, muscle spasms, heart palpitations and breaks up mucous congestion especially when taken with cayenne pepper.

b. It is also used to calm nerves and to lower the blood pressure.

c. This mint when simmered in an open pot will sweeten, disinfect and germicide the air.

d. As a bath it will relieve stress and fatigue, bodily aches and pains.

e. When cooked with dried peas and beans it neutralizes gases and acids, and makes them easy to digest.

f. It keeps the body warm in damp and wet weather.

PERIWINKLE (Vinca Rosea)

Some common names for periwinkle are little pinkie, bright eyes, old maid, ram goat roses and brown man's fancy.

Medicinal and other uses:

a. A tea made of this herb is used for diabetes and the white flowering plant is used for high blood pressure.

b. Chewing the fresh herb will relieve toothaches.

c. It is also used to stop excessive menstruation and to stop bleeding in the nose and mouth.

PIMENTO (Pimento officinalis)

Some common names for pimento are allspice and clove pepper.

Medicinal and other uses:

a. Pimento leaves held in the mouth will decrease the urge for cigarettes.

b. In Jamaica, a tea of the leaves is used for chills, pneumonia and flu, and also to improve circulation.

c. As a spice or seasoning it acts as a natural preservative to prevent foods from spoiling.

d. A tea of the leaves will also keep the body warm, purify the stomach, aid digestion, relieve gas and stop vomiting.

e. A mixture of pimento and white rum is an excellent remedy for menstrual cramps

PLANTAIN (Plantago Major)

Some common names for plantain are lante, common plantain and white man's foot. The plant is rich in healing chlorophyll and potassium. The fruit is edible and can be prepared in different ways.

Medicinal and other uses:

a. In Jamaica, a tea of the leaves is left out overnight to collect dew and is then used to stop diarrhea and internal bleeding and to remove dysentery.

b. A poultice of the fresh leaves is applied to swellings, boils and infections.

c. A tea of the entire plant can be used for bladder and kidney problems, stomach ulcers, hay fever and allergies.

d. The early morning dew, collected from the leaves, can be applied to the eyes to strengthen them, as well as to the skin for all dermatological problems and pain.

e. The fresh leaves placed in one's shoes relieves foot pains and fatigue.

f The fresh leaves are also used to dress wounds, dog and insect bites, to stop bleeding and prevent infections.

g. The juice extracted from the leaves, is used as an eyewash for those suffering from opthlamia.

POLICE MACKA (Tribulus Cistoides)

Medicinal and other uses:

a. Tea made from this plant can be taken for kidney and bladder infections, colds and malaria.

PRICKLY POPPY (Argemone Mexicana)

Some common names for prickly poppy are yellow thistle and spiritweed.

Medicinal and other uses:

a. In Jamaica, a tea made of the flower is used for fevers and colds, especially in infants.

b. The sap can be used externally for ringworms, warts, ulcerated wounds and sores.

RICE BITTERS (Androgrhis Paniculata)

Medicinal and other uses:

a. Rice bitters are used in the Caribbean for colds, mucous congestion, fevers, nerves, cholera and dysentery.

ROSEMARY
(Rosmarinus Officinalis)

There are two basic kinds of rosemary. One having a blue flower, the other white, both having the same properties. The blue species is erect while the white is creeping. The blue is the most commonly used of the two.

Medicinal and other uses:

a. Rosemary is used to soothe and calm the nerves, relieve mental fatigue and headaches, especially migraines.

b. Its disinfectant properties, makes it very useful for gum and other mouth problems.

c. In the Jamaican countryside, rosemary is smoked to relieve nasal congestion, asthma attacks and chest congestion.

d. It is also simmered in an open pot and the vapours are inhaled as a highly effective relief for hay fever and allergies.

e. When used in food preparations, it not only flavours but also purifies food by acting as a natural preservative to prevent souring or spoiling.

f. It also neutralizes gases and aids digestion.

g. Mopping the floor with rosemary keeps away flies and other insects.
h. As a beauty aid, rosemary makes one of the finest hair and scalp rinses.
i. It cures dandruff, and other scalp ailments, strengthens hair, and with regular use it will restore natural colour to hair.
j. Rosemary rinse will also prevent curls from uncurling in damp weather.

SALVIA

Medicinal and other uses:

a. Used as a bath for bodily aches and pains.

SAINT JOHNS WORT (Hypericum Perforatum)

Medicinal and other uses:

a. A great remedy for injured nerves especially the nerves of the toes and fingers that have been crushed.
b. It also prevents lockjaw; and relieves pain-after operations.

SARSAPARILLA (Smilax Officinalis)

Some common names for sarsaparilla are reed grass and Jamaican sarsaparilla. There are many species of sarsaparilla in colours ranging from gray, brown, orange and red.

Medicinal and other uses:

a. Sarsaparilla is one of the best blood purifiers and general tonics.

b. It is very good at breaking up infections in the body by eliminating wastes through urine and perspiration.

c. It is effective for all skin ailments such as eczema, psoriasis, ringworm, arthritis and rheumatism.

SASSAFRAS (Sassafras Albidum)

Medicinal and other uses:

a. Sassafras tea is used to reduce flesh and move fat and mucous from the body.

b. The bark added to dried fruits and grains will prevent them from spoiling, and keep out insects.

c. The bruised leaves can be used as a poultice for pains, sprains and sores.

SEA GRAPES
(Coccaloba Urifera)

Medicinal and other uses:

a. The fruit is very nutritious reduces fever and keeps the body cool.

SEMI CONTRAC
(Chenopodium Ambrosiodes)

Some common names are West Indian goose foot, Worm grass and American wormseed.

Medicinal and other uses:

a. In Jamaica, the tea is used as a laxative to clean the intestines, remove wastes and parasites from the stomach.

b. The tea is also used as a remedy for colds and asthma.

c. Suppositories are made by powdering the dried leaves along with spearmint, salt and water. This mixture is shaped and inserted into the rectum for appendicitis.

d. This is one of the best herbs for intestinal worms, especially in children.

SHAME OLD LADY (Mimosa Pudica)

Medicinal and other uses:

a. Tea of entire plant is used as an antidote for poisons, a sedative, venereal disease and nerve problems.

SORGHUM BICOLOR

Some common names are chicken corn, Creole corn and Jerusalem corn. Sorghum is rich in protein, B vitamins, calcium, very high in nitrogen; also contains other minerals and vitamins, fiber and carbohydrates in the form of unconcentrated starch.

Medicinal and other uses:

a. The seeds are ground into sorghum meal, or flour, which is used to make bread tortillas, porridge and other dishes.

b. The straw is used to make baskets and house roofs.

SOURSOP (Annona Muricata)

Soursop is rich in vitamin B, C, calcium, and minerals. The fruit of the soursop has a leathery, somewhat spiky outer skin and the inside looks somewhat like wet cotton. It can be eaten fresh or made into soursop nectar or dairy free ice-cream.

Medicinal and other uses:

a. In Jamaica the leaves are used for kidney and gall bladder problems and to eliminate inorganic calcium deposits in the joints.

b. The crushed leaves are used as a poultice for wounds and sores.

c. In the Caribbean the leaves are used to make a common bush tea which is taken daily by both children and adults as a nerve tonic.

d. A tea of the dried fruit is commonly used for dysentery and jaundice.

e. The crushed leaves can be used to revive one from a fainting spell.

f. A tea made of the leaves is effective against worms.

SPANISH NEEDLE (Bidens Pilosa)

Medicinal and other uses:

a. A tea of the flowers is used for angina and other heart problems.

b. The tea also internally keeps blood cool and can be used for diabetes.

SWEETSOP (Annona Squamosa)

The fruit is rich in vitamins and minerals and is quite delicious.

Medicinal and other uses:

a. The crushed leaves can be applied to malignant sores and ulcers.

b. The aroma of the crushed leaves will prevent fainting and the leaves can be tied around the head to prevent headaches.

c. A tea of the leaves can be taken for pains in the spleen and for pain during childbirth.

TAMARIND (Tamarindus Indica)

The ripe fruit contains 10% tartaric acid, natural acetic, malic, also pectin and natural sugar, vitamins and minerals. The pulp of the fruit is made into pudding and a candy known as tamarind balls.

Tamarind

Medicinal and other uses:

a. In Jamaica a tea of the leaves is used for measles, as a bath for fevers, smallpox, chicken pox and bodily pain.

b. The leaves can also be used as an eyewash for red swollen eyes.

c. The roots can be made into a tea for hemorrhages.

d. A tea of the leaves can also be taken for chest colds, coughs, throat irritations and diabetes.

TOBACCO LEAF

Medicinal and other uses:

a. The cured leaves is used in tincture externally for all bodily aches and pains.

b. A tea made of the leaves makes an effective insecticide when sprayed on crops.

c. Tobacco leaves used as a poultice draws poisons and infections from boils and sores.

TRUMPET TREE (Cecropia Peltata)

Some common names are wild pawpaw, snakewood and guarina. The young inner buds are a very nutritious vegetable when steamed.

Medicinal and other uses:

a. The sap of the tree is used to make an ointment for skin problems and to make rubber.

b. A tea of the young shoots can be used for toothaches, whooping cough, fever and as a heart tonic.

c. In Jamaica the tea is used for asthma, flu, mucous congestion, hoarseness and to expel the placenta after childbirth.

THYME
(Thymus Vulgaris)

Thyme is one of the tastiest herbs to use in the kitchen. It imparts a lively fresh taste to all dishes. Thyme honey is one of the sweetest in the world.

Medicinal and other uses:

a. Thyme is soothing to the nerves and helps to cure insomnia.

b. Thyme tea is one of the best cough remedies.

c. It brings almost instant relief to whooping cough and is very effective for asthma, shortness of breath, lung problems and bronchitis.

d. Thyme is one of the best blood purifiers. It stimulates the flow of blood to the skin surface.

e. Thyme baths and hot compresses are an excellent remedy for pains, bringing relief for headaches, labour pains, arthritis pain, pulled muscles, sprains and stomach cramping.

TUNA OR PRICKLYPEAR (Opuntia Ficusindia)

Some common names are Indian figs and opuntia. It is available as a house plant at most local florists. The fruit is edible and quite nutritious, and the young leaves can be steamed and used as a vegetable.

Medicinal and other uses:

a. In Jamaica the leaves are sliced open and soaked overnight and taken to relieve constipation, to clean the intestines, bring down fevers and remove mucous from the body.

b. A tea of the leaves is taken to relieve excessive menstruation and is mixed with cerasee and taken daily to relieve backaches.

c. A tea of the root is used to combat dysentery and gonorrhea.

d. The fresh sap of the leaves is used as shampoo to clean and strengthen the scalp and hair, eliminating dandruff, itching and other scalp problems.

WOMAN PIABA BUSH (Hyptis Pectinata)

Medicinal and other uses:

a. Used in Jamaica to relieve pains during child birth, for stomach pains, fever and chest colds.

b. Fresh juice of this plant is used for sores and wounds.

Herbal Myths & Legends

JAMAICAN WILD SWEET BASIL

In Jamaica, it is believed that sweet basil, when planted by the doors, gates and in window boxes, will prevent evil or negative vibrations from entering the household.

KHUS KHUS

It is believed that if Khus Khus is carried around it will attract positive people, or if placed about the household will attract good vibrations.

LEAF of LIFE

Legend has it that the leaves placed over the door of your loved one will tell you if your lover has been faithful. If a new plant grows from each notch, your lover has been loyal. The amount of notches that do not produce a new plant indicates the number of times your loved one has been unfaithful.

MANDRAKE ROOT

Carry a piece in your pocket or pocketbook where you carry your money and you will never go broke.

PAPAYA

It is believed that if a young man beats his penis on the trunk of a papaya tree it will grow to a large size when he gets older.

PEPPERMINT

When simmered in an open pot will bring harmony to the household.

SALVIA

Believed to cleanse and drive evil forces and bad luck.

SPIRIT WEED

When picked during certain phases of the moon, it has the power to make the person who chews the root invisible. The Maroons of Jamaica used it against the British during their resistance, and reports from British soldiers state that they could not see the enemy, throwing away their weapons they ran back to their camps with reports that the trees were fighting them, because they could see the leaves moving and hear the rustling but could see no one.

STAR ANISE

Will drive away misfortune from household and attract good fortune if placed in a saucer on a table.

YARROW

If a piece of yarrow is sewn into a cloth bag made of cotton or flannel and put under the pillow, one would see future sweethearts or the one to marry, in a dream.

Illnesses and Treatments by Plants

ARTHRITIS

Rue, cherries, grapefruits, dandelion root tea, sarsaparilla root tea and celery seed tea.

ASTHMA

Comfrey root juice and leaves, lemon tea, horse-radish and lemon juice and nettle.

BRONCHITIS

Pimento leaf, rosemary, sweet basil ginger, nutmeg and cloves.

BACKACHE

Nettle.

BLADDER PROBLEMS

Flaxseeds and watermelon.

BLOOD PRESSURE

[Lowering] A tea made from dried watermelon seeds will lower high blood pressure. So will

breadfruit blossom when steamed and also cho-cho juice.
[Raising] Rosemary, garlic and cayenne pepper can raise blood pressure.

BLOOD PURIFIERS

Dandelion and sarsaparilla roots.

BODILY ACHES & PAINS

Sage, pepper elder, pimento leaves or seeds.

BONES

Comfrey root is good for diseased bones.

BREATH

Leaf of life, eucalyptus leaves, thyme and peppermint are all breath fresheners.

BURNS

If there are no blisters, apply honey. Marigold flowers relieve burn pains and prevents scars from forming. Fresh aloe also prevents scars and blisters.

CANCER

Sarsaparilla, sassafras root, mandrake and marigold flowers.

CIRCULATORY SYSTEM

Herbs helps one to have a healthy heart, veins and normal blood pressure.

CORNS

Lemon juice, garlic, onion, pineapple- apply directly on corn, tape closed.

COUGHS

Peppermint tea or strong thyme tea mixed with honey.

DANDRUFF

Rosemary rinse, nettle, chamomile, aloe vera and cannabis are all excellent for dandruff.

DEODORANTS

Apply lemon grass, sage, fresh parsley and chamomile.

DIABETES

Dandelion root tea, mango leaves, periwinkle, cerasee, eucalyptus leaf, Spanish needles and lignum vitae leaves.

DIARRHOEA

Nettle and peppermint.

DIGESTION PROBLEMS

Lemon juice, Jamaican wild sweet basil and cola nut.

DIGESTIVE SYSTEM

Good health begins here and herbs assist the digestive system to be in good working order.

EAR INFECTION

Warmed gel of the Aloe Vera dropped into the ears will get rid of infections.

EXHAUSTION

Passionflower.

EYES

Rosemary tea and passionflower used as an eyewash.

FAINTING

Inhale the aroma of freshly crushed soursop leaves.

FEVERS

Fever grass, Spanish needle, soursop leaf, chamomile and mint tea.

Putting the afflicated individual to lay in a bed lined liberally with leaves of the Guinea Hen Weed, will help in getting rid of the fever.

FLEA & LICE REPELLANT

Teas made of pimento leaf, sage, castor or coconut oil can be applied to the body externally.

GANGRENE

Apply chamomile poultice to wound.

GLANDULAR SYSTEM

Herbs promote a healthy sex drive, helps in coping with menopause and PMS.

GUMS

Comfrey root when used as a mouthwash strengthens gums and stops bleeding.

HAIR

Aloe vera, cannabis, sage, nettle and chamomile.

HEADACHES

Chamomile, peppermint, rosemary and orange juice.

HEART

Peppermint and chamomile are good for the heart.

HEART BURNS

Chew ginger, corn or suck on a fresh lemon.

HEMORRHOIDS

[Internal bleeding] Marigold flowers.

IMMUNE SYSTEM

Herbs strengthen the body's defenses against disease.

INFECTIONS

Sarsaparilla root, garlic and cayenne pepper.

ITCHING

Spanish needle or tamarind tea taken internally or externally.

JOINTS

Nutmeg tincture, pepper elder bath.

LUNG PROBLEMS

Comfrey root, marigold flowers, periwinkle and nettle.

MEASLES

Sassafras.

MENSTRUATION PAINS

Pepper elder and pimento leaf

MENOPAUSE

Passionflowers.

MINERAL DEFICIENCY

Sarsaparilla root and Irish moss.

MEMORY & BRAIN

Rosemary and chamomile.

MUCOUS CONGESTION

Irish moss, black sage, ackee, pepper elder and peppermint.

NERVES

Passionflower and mint are very soothing to the nerves.

NERVOUS SYSTEM

Herbs assist in coping with stress & insomnia.

NURSING MOTHERS

Anise, nettle and dill increases the flow of breast milk.

OPEN BREATHING PASSAGES

Rosemary and cannabis.

PNUEMONIA

Pimento leaf and eucalyptus.

PREGNANCY

Nettle leaves relieves morning sickness

PROSTATE STRENGTHENER

Irish moss tea.

POISONS [ANTIDOTES]

Cola nut and flaxweed.

RESPIRATORY SYSTEM

Breathing problems can be stemmed by the use of herbs.

SKELETAL SYSTEM

Herbs promote strong bones and joints.

SMOKING HABIT

Chewing on a Jamaican chewstick will lessen the urge for cigarettes and eventually eliminate the habit.

SOAPS & SHAMPOOS

Nettle, rosemary and chamomile.

SPLEEN

Dandelion root tea.

STOMACH

Nettle, chamomile and cloves.

SUNBURN

Spanish needle, nettle, sage and tamarind leaves can be applied to skin.

SWELLING

Thyme, marigold flower and flaxseeds.

SYPHILIS

Sarsaparilla root tea and lignum vitae.

TEETH

Comfrey root tea used as mouthwash will prevent tooth decay.

Powdered nutmeg can be stuffed into cavities.

TUBERCULOSIS

Irish moss, flaxseed and cloves.

URINARY PROBLEMS

Nettle, linseed and parsley.

URINARY SYSTEM

Herbs strengthens and cleanses the urinary system.

VOMITING

Marigold flowers tea prevents recurring vomiting.

WATER RETENTION

Irish moss and dandelion root tea.

WHO NEEDS HERBS?

Anyone who is desirous of being in good health and maintaining a healthy lifestyle should incorporate the use of herbs in their daily lives. For people who engage in strenuous activity such as athletes and body-builders herbs are the perfect complement.

Children and babies need herbs too and should not be left out. Parents and guardians should ensure that their children enjoy the many benefits of herbal use.

HERBS REALLY GET THE JOB DONE!

There are several facts about herbs that are beneficial for one to know.

a) Herbs feed, regulate and cleanse the body naturally.

b) Herbs will not build up in the body or produce harmful side effects like synthetic drugs.

c) Herbs were actually created to root out the cause of disease.

Herbs may be used singularly or in a combination. These combinations are herb food formulas composed of two or more herbs to aid a certain system. An herbal combination is chosen to specifically address the entire complaint of an individual. The herbs that best address their particular symptoms are chosen over similar plants.

Several plants or their extracts can work together in a balanced fashion.

What one herb lacks another can provide, so that the combined action improves what can be accomplished by a single herb. Some herbs in the combination would help relieve the symptoms while others act to correct the cause of the symptoms.

HERBAL QUALITY

Herbal quality cannot be over emphasized. It is critical to effective herbal therapy that the proper plants are picked in the proper season and used fresh. High quality herbs will retain all the characteristics of the whole herb: aroma, colour, taste and effect.

Whenever possible the plants used should be organically grown or locally abundant herbs can be specifically wildcrafted to avoid contamination. Many commercially available bulk herbs contain residues from agricultural chemicals, fumigation and irradiation.

Organic cultivation allows the manufacturer of herbal extracts to maintain access to high quality botanical ingredients. A recent advance in herb technology and research (fresh freeze-drying) allows maintenance of the natural potency of the herbs by preserving most of the biologically active constituents of the fresh plant. In many instances, improved or unique therapeutic action has resulted from the fresh freeze-drying process.

HERBAL EXTRACTS

Herbal extracts have been used in many forms and strengths as galenicals, tinctures, fluid extracts, etc. These are water and alcohol extractions made from fresh or shade-dried plants. Some extracts include the addition of a little vegetable glycerin. A few herbs are also extracted in 100% organic olive oil for external use.

Herbal extracts offer the advantage of being more readily available to the body than powdered herbs. These plant extracts are effective preparations which are well tolerated. They may be taken alone or in a little water or juice.

HERB USAGE GUIDE

- Immediately stop taking any herb if unusual symptoms develop.
- Consult your health care provider concerning the use of herbs in pregnancy or with infants.
- Some herbs, in large doses, may upset stomach or loosen the stool.
- Driving and potentially dangerous activity is discouraged when utilizing sedative herbs.
- Long term use of diurectics without monitoring may be dangerous.
- Many herbs work faster on an empty stomach, but if they produce nausea, they should be taken with a light meal.
- Combining herbs may increase or change the expected action of each individually.
- Herbs are not generally recommended as first aid, emergency-type medicines.
- Chronic, severe or worsening symptoms should be evaluated by a licensed health care provider.

Herbs of the World

ALFALFA - Nutritive, with high mineral and vitamin content, including vitamin K and iron; contains phytoestrogen precursor for menopause; lowers cholesterol.

ANGELICA - For strong menstrual cramps with scanty flow; for intestinal colic, poor digestion; stimulating expectorant for coughs.

ARNICA OIL - EXTERNAL USE ONLY - For bruises, swelling and other athletic injuries.

ASTRAGALUS - Primarily promotes and stimulates immune function. Mild antibiotic; enhances endurance.

BAYBERRY- Astringent for venous congestion, especially in respiratory or GI mucosa, as gargle for sore gums or sore throat.

*BLACK COHOSH - Female tonic, relaxes, smooth muscle, relieves nervous irritation and menstrual cramps; reduces joint inflammation as in arthritis.

BLACK WALNUT-Astringent, antiseptic, purifier, especially for skin eruptions; anti fungal and vermifuge in larger doses.

BLOODROOT - Digestive tonic, bronchial expectorant; topically for skin infections, as gargle for oral lesions (prevents plaque build up).

*BLADDERWRACK - Supplies trace minerals, good for low thyroid, obesity and removing toxic minerals.

*BLUE COHOSH - Female tonic, improves uterine tone.

BLUE VERVAIN - Sedative and diaphoretic; useful in early stages of flu.

BUCHU - Diuretic and urinary antiseptic; for gravel, infection of the bladder, acid urine.

*BURDOCK - Blood and liver cleanser; for skin eruptions; gout, diabetes; mild immune stimulant.

BUTTERNUT BARK - laxative and mild anti parasitic (similar to Black Walnut).

*CALENDULA - For infections, colds; mild immune and circulatory stimulation; excellent topically for skin injuries or lesions, including burns.

CASCARA SAGRADA - Laxative, best used sporadically in chronic constipation.

CATNIP - Children's tonic, colic, teething, also any stomach upset, gastritis.

*CAYENNE - For first stages of a cold when skin and mucosa are hot and dry; a stimulant, improves circulation; aids utilization of other herbs; stimulates digestion and appetite; helpful in asthma formulas.

*CELANDINE - For biliary colic, chronic liver disease, jaundice good overall liver tonic and cleanse; for skin eruptions.

CHAMOMILE - A calmative to gastrointestinal and nervous system, especially in children (colic, teething), and the elderly (improves appetite, digestion and sleep).

CHASTE TREE (Vitex) - Menstrual regulator, improves circulation and tone of female organs, promotes progesterone production. CHERRY BARK - Tonic and expectorant, especially in bronchial congestion.

CHICKWEED - Mild diuretic for water retention; demulcent for coughs; topically for skin eruptions and hemorrhoids.

CLEAVERS - Good diuretic, blood purifier acts to reduce gravel or stones.

COLTSEOOT - Demulcent for bronchial and gastric irritations including asthma, cough and lung congestion.

CORNSILK - Diuretic; may be used for bladder and prostate problems.

CRAMP BARK - Antispasmodic for dysmenorrhea and other rhythmic uterine pains; for smooth muscle cramps or spasms including intestinal; sometimes used for threatened miscarriage.

*CRANBERRY - Strong diuretic and urinary antiseptic (acidifies the urine).

DAMIANA - Mild diuretic and tonic for nervous depression and poor appetite; for recurring genitourinary complaints with emotional causes.

*DANDELION - Excellent diuretic (leaf) and liver cleanser (root); lowers cholesterol and blood pressure.

DEVIL'S CLAW - Anti-inflammatory, possibly acting on prostaglandin pathways; useful in arthritis and other join inflammations.

DEVIL'S CLUB - Blood purifier; builds endurance; has been used for lowering blood sugar.

DONG QUAI - Menstrual regulator; the Chinese say it "regulates the hormonal balance and flow."

*ECHINACEA - Strong immune stimulating properties; useful in colds, flu, sore throat, infections, skin eruptions, allergies, viral disease and immune deficiency.

ELDER FLOWERS - Expectorant, diaphoretic (causes sweating); blood purifier, and mild laxative; long traditional use for early stages of coughs, colds and flu.

EPHEDRA - A strong nerve stimulant, relieves swelling of mucus membranes and opens air passages due to antispasmodic action; used in asthma and allergies.

EYEBRIGHT - For colds, sinus congestion, and allergies; acts as topical astringent; for mild eye inflammations - use when there are watery discharges from nose and eyes.

FENNEL - Aids digestion, gas and appetite; used with laxatives to prevent cramping.

*FEVERFEW - For headache or rheumatic inflammation; prevention of migraine headache (only works in the freeze-dried form.

*GARLIC - For general health enhancement; strong antibiotic with antimicrobial, antifungal and anti-parasitic action; lowers blood pressure; protects against colon cancer; a stimulant, diaphoretic, expectorant and stomachic.

GENTIAN - A bitter tonic and stomachic; stimulates appetite and digestive secretions; especially useful with debilitated persons.

*GINGER - For indigestion and gas pain; motion sickness; a general stimulant and aids respiratory congestion; joint pain.

*GINKGO - Improves peripheral circulation; helpful in memory, hearing and potency restoration and improvement of cerebral and coronary blood flow.

GINSENG - Used for nervous disorders and indigestion; a stimulant for mental, physical and nervous exhaustion; aids utilization of food and serves as a nutrient. Adaptogen for stress. Normalizes body systems such as blood sugar; male tonic.

*GOLDENSEAL - For indigestion, and appetite stimulation; combats colds, diarrhea, flu and infection; use glycerin extract topically for skin ulcerations and mucous membrane inflammations.

GOTU KOLA - For emotionally caused depression with sluggish digestion; poor circulation and low endurance and energy; stimulant; promotes mental clarity; topically for minor skin abrasions.

GRAVEL ROOT - Soothes mucosal irritations - especially in urinary tract, bladder and prostate; flushes sediment from kidney.

*HAWTHORN - Strong heart tonic (strengthens heart muscle and improves coronary blood flow); lowers blood pressure.

*HOPS - For gastric spasms due to nervousness; strong muscle relaxant; sedative for anxiety; promotes restful sleep; estrogenic properties.

HOREHOUND - An excellent expectorant for coughs and lung congestion.

*HORSETAIL - Diuretic and astringent, useful in gravel and kidney problems. High mineral content includes silica and calcium.

HUCKLEBERRY - Astringent, containing tannins and anthocyanins; improves vision; stimulates circulation, especially in venous system; useful with diarrhea, eye disorders, hemorrhoids and sugar metabolism problems.

HYSSOP - Carminative for indigestion/gas; also a diuretic, mild respiratory sedative; expectorant.

*JUNIPER BERRY - Diuretic, useful in bladder or kidney conditions, especially chronic bladder infections.

LICORICE - A demulcent and expectorant - for coughs and respiratory congestion; stimulant; promotes estrogen; improves adrenal function.

*LOBELIA - Strong bronchial dilator, expectorant, antispasmodic and larger doses, emetic; use for asthma and any lung congestion, hiccoughs; helpful in stopping smoking.

*LOMATIUM - Antibacterial and immune stimulating; for colds, flu, viral sore throats, chronic viral disease, respiratory infections and congestion; antifungal and anti-parasitic.

MARSHMALLOW - Excellent demulcent for soothing mucous membrane irritation in gastrointestinal, respiratory and urinary tract disease.

MILK THISTLE - For liver disease and cleansing; actually aids liver regeneration after toxic exposure.

MOTHERWORT - Antispasmodic and female tonic (improves blood flow to the female organs); sedative for hysterical complaints, tachycardia, nervous pulse.

MULLEIN - Demulcent, expectorant and anodyne properties make it useful with irritations of the lungs and bowels. The oil can be used for ear problems.

MYRRH - Soothes mucous membrane irritation; astringent, stimulant, stomachic; good topically for wounds or as gargle for oral lesions; immune stimulating and antifungal.

*NETTLES - An astringent and alkalizing diuretic; good nutritive and high mineral content; a mucosal astringent that decreases secretions.

*OAT - Nervine, stimulant and antispasmodic; nutritive to nervous system; restores energy and vital force to weakened system; helpful in breaking addictions.

OCOTILLO - Relieves pelvic lymphatic congestion such as "boggy" uterus and prostate.

OREGON GRAPE - Liver tonic and cleanser, blood purifier; for poor protein digestion or hangovers; for sluggish gastrointestinal system.

*OSHA - Aid for bronchial irritations, congestion, or coughs that are dry; anti viral, immune stimulating properties.

*PASSION FLOWER - antispasmodic and nerve sedative; good for worry, anxiety or depression, especially menstrual, or in children or the elderly; induces restful sleep; aid for neuralgia, tense muscles, dysmenorrhea and hypertension.

PAU D'ARCO - Antiseptic and blood purifier; used with candida, lymph congestion, and tumors; improves GI utilization of nutrients.

PEPPERMINT - Relieves gas pains, aids digestion, calms nausea and diarrhea as in flu.

PIPSISSEWA - Diuretic, astringent, excellent for urinary tract infections, especially if kidneys involved including gravel and stones; use when urination is scanty, painful and contains mucous.

PLANTAIN - Diuretic, antiseptic, and astringent, used topically for wounds in healing salves or internally for mucosal irritations.

PLEURISY ROOT - Antispasmodic, diaphoretic, expectorant; for bronchitis, pneumonitis, pleurisy, especially with hot, dry mucosa.

PRICKLY ASH - Stimulant and astringent for poor peripheral circulation or venous congestion including pelvic, also mouth ulcerations.

PROPOLIS - Antiseptic with soothing and healing properties; good for mouth, gum and skin ulcerations.

*RED CLOVER - Strong blood purifier; nutritive to nervous system; aid for spasmodic cough, ulcerations and tumors.

*RED RASPBERRY - Relieves nausea and improves digestion; improves uterine tone and blood supply; increases milk production.

RED ROOT - Expectorant astringent antispasmodic for

tonsillitis and bronchitis or any lymph swelling, also with enlarged spleen or liver; hastens clotting time.

RHUBARB - Appetite stimulant in small doses, and tonic bitter for indigestion and gas; in large doses a laxative/purgative.

SARSAPARILLA - Soothes mucous membranes; relieves gas; useful with rheumatism and serves as a precursor for hormone production; excellent digestive tonic.

*SAW PALMETTO - Diuretic; enhances endurance; breaks down mucous in respiratory congestion; useful in prostate (BPH), urethritis and kidney problems.

SHEPHERD'S PURSE - Hemostatic astringent; helps stop passive gastrointestinal or uterine bleeding; diuretic, breaks down urate stones.

*SHIITAKE MUSHROOM - Stimulates the immune system aiding response to chronic infection, tumors, and debilitated states (Fresh freeze-drying uniquely preserves the active polysaccharides).

*SKULLCAP - Antispasmodic and nervine; for nervous muscle activity (spasms or twitches); for insomnia, anxiety and neuralgia.

SLIPPERY ELM - For indigestion and gas; soothes mucous membrane in GI and respiratory disease.

*St JOHN'S WORTH - For agitation and depression; mild immune stimulation and antiviral; used topically over migratory or shooting pains, especially with nerve involvement. May aid with children's bed wetting.

STONE ROOT - Sedative, antispasmodic, astringent, diuretic; for mucous or gravel in the urine, or any pelvic venous and lymphatic congestion including, hemorrhoids, swollen prostrate laryngitis.

USNEA - Antibiotic; useful in bacterial infections (especially of the intestine), and colds and flu where it also has immune stimulating properties and in fungal conditions.

*UVA URSI - Good diuretic, for urinary tract disorders, gravel in the urine (potency enhanced in freeze-dried form).

*VALERIAN - Strong nervous system sedative; for anxiety, irritability, insomnia due to excessive worry or depression; mild pain relief.

WHITE WILLOW - Tonic and astringent (contains tannin and salicin); brings down fevers, aid for indigestion due to disease; may relieve headache or joint inflammation.

WILD YAM - Antispasmodic; relieves colic; spasm or cramping of smooth muscle including gall bladder and uterus; hormone precursor.

WITCH HAZEL - Astringent, tonic, sedative for mild hemorrhage; topically for varicosities and hemorrhoids.

WORMWOOD - Bitter tonic, stomachic for feeble digestion; brings down fevers; clears the intestines of worms.

*YARROW - Diaphoretic, stimulant; for colds and fever; slows passive bleeding from boggy mucosa.

*YELLOW DOCK - Blood cleanser (high iron content); helpful in anemia, chronic skin disorders and congestion; for poor fat digestion and constipation.

YERBA MANSA - Improves lymph drainage in mild colds, sore throats and sinus infections; also in subacute colitis and cystitis.

YERBA SANTA - Expectorant and antispasmodic; for coughs, bronchitis and asthma; anti fungal.

YUCCA - Contains steroidal saponins including sarsapogenin, a potential cortisone precursor; used internally and externally for slow healing ulcers and skin eruptions, arthritis and joint inflammation.

Published by
Eclectic Medical Publications,
Saudy, Oregan

*Denotes herbs which are preferably used in the fresh freeze dried form.